HEALTHY ME

Dairy Products

HEALTHY ME

Published by Smart Apple Media
1980 Lookout Drive, North Mankato, Minnesota 56003

PHOTOGRAPHS BY Heartland Images (Paul T. McMahon), Bonnie Sue Rauch,
Tom Stack & Associates (Tom McCarthy, Brian Parker, Robin Rudd), Unicorn Stock Photos
(Eric R. Berndt, Andre Jenny, Gary Randal, Aneal E. Vohra)
DESIGN BY Evansday Design

Library of Congress Cataloging-in-Publication Data
Kalz, Jill.
Dairy products / by Jill Kalz.
p. cm. — (Healthy me)
Includes bibliographical references.
Summary: Describes various dairy products and their role in human nutrition.
Includes instructions for making butter.
ISBN 1-58340-297-7
1. Milk—Juvenile. 2. Cookery (Milk)—Juvenile literature.
3. Dairy products—Juvenile literature. [1. Milk. 2. Dairy products. 3. Nutrition.] I. Title.

TX556.M5 K35 2003
641.3′71—dc21   2002030625

First Edition

9 8 7 6 5 4 3 2 1

Dairy
Products

Dairy Cows

Some cows are brown. Others are black and white. Some cows live in the mountains. Others live by the sea. There are many different kinds of **dairy** cows, but all of them make the same thing. Milk.

A cow makes milk to feed her babies. She keeps

the milk in her **udder**. A farmer gets the milk by

pulling down on the cow's **teats**. He does this

with his hands or with machines.

5

Squeezing milk from a cow is called milking. ⌃

< A group of cows is called a herd.

Trucks take the milk to the dairy in cold, steel tanks. At the dairy, the milk is tested. It is heated and cooled to kill germs. Then it is put into cartons or bottles and sent to stores.

In some parts of the world, farmers get milk from goats, sheep, camels, or alpacas.

Farmers send milk to the dairy in trucks.

Made from Milk

Raw milk is milk that has just come out of a cow. When raw milk sits, a layer of fat rises to the top. This fat is called cream. A dairy mixes the cream into the milk. Whole milk has a lot of cream in it. Skim milk has just a little.

Milk and foods made from milk are called dairy products. Milk is used to make yogurt and cheese. Cream that is beaten for a long time turns into butter. Most ice cream is made with milk. So is sherbet.

Most ice cream has milk in it.

Many farmers use machines to milk cows.

The most common kind of dairy cow in
North America is the Holstein.
It is black and white.

Wisconsin has the most cows in
the United States. But California
makes the most milk.

The Good Stuff

Dairy products have important **vitamins** and **minerals** that everybody needs. Vitamin D and calcium build strong bones and teeth. Without calcium, bones break easily.

Dairy products also have vitamin A and protein.

Vitamin A keeps hair, skin, and eyes healthy.

Protein builds strong muscles.

Dairy products help your body grow. ⌃

‹ Healthy bodies have strong bones and muscles.

Not all foods made from milk are good for you. Most ice cream has sugar added. Most cheese has a lot of fat. Too much sugar and fat can be bad for you. Small helpings of these foods are okay. But drinking milk is the best way to get your vitamins and minerals.

An average dairy cow makes enough milk in one day to fill 90 glasses.

Ice cream has very few vitamins or minerals.

Eating Right

All foods belong to one of five food groups.

Foods made from milk belong to the dairy group.

Fruits belong to the fruits group. There are also

groups for vegetables, meats, and grains.

Doctors say you should eat or drink three helpings from the dairy group each day. A helping may be a glass of milk. A small dish of yogurt. Or two slices of cheese.

Yogurt comes in many different flavors. ⌃

< Your body needs all kinds of foods.

17

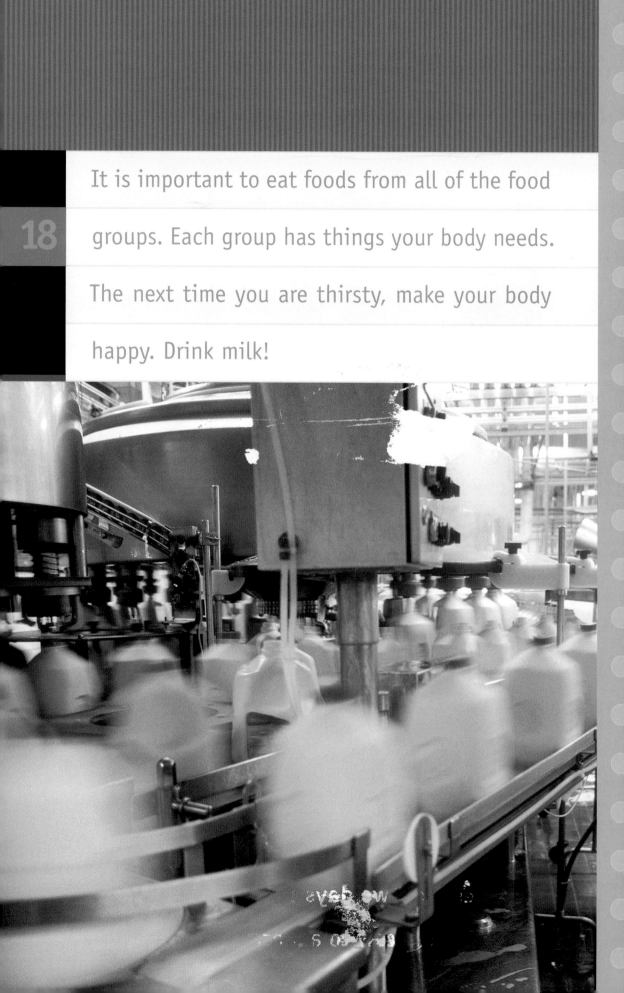

It is important to eat foods from all of the food groups. Each group has things your body needs. The next time you are thirsty, make your body happy. Drink milk!

18

Chocolate milk does not come from brown cows. It is just white milk flavored with chocolate.

It takes just two days for milk to go from a cow to a store.

20

Making Butter

No butter in the refrigerator? Have an adult help you make some!

WHAT YOU NEED
One-half pint (240 ml) heavy cream
A mixing bowl
An electric mixer

WHAT YOU DO
1. Pour the cream into the bowl.
2. Have an adult help you turn the mixer on "high." After a few minutes, the cream will turn into whipped cream. Keep beating!
3. Turn the mixer off when the cream looks watery. This "water" is called buttermilk. The yellow lumps you see are butter.
4. Pour out some buttermilk. Then spread the butter on a piece of bread. Delicious!

dairy milk or milk products; also, a place that handles milk and milk products

minerals things in food that keep your body healthy and growing; calcium is a mineral

teats nipples on a cow's udder; the cow's milk comes out through them

udder a cow's "milk bag"; it hangs down from the cow's belly

vitamins things in food that keep your body healthy and growing

Read More

Cooper, Jason. *Dairy Products*. Vero Beach, Fla.: Rourke Publications, 1997.

Gibbons, Gail. *The Milk Makers*. New York: Simon & Schuster Children's Books, 1986.

Llewellyn, Claire. *Milk*. New York: Scholastic Library Publishing, 1998.

Explore the Web

THE "GOT MILK?" CAMPAIGN

http://www.whymilk.com

http://www.got-milk.com

MOOMILK: A DYNAMIC ADVENTURE INTO THE DAIRY INDUSTRY

http://www.moomilk.com

POWERFUL BONES, POWERFUL GIRLS

http://www.cdc.gov/powerfulbones

Baby cows drink milk from their mothers' teats.